The Collector's Series / Volume 37

The wonderful BERRY BOOK

Delicious Recipes

by Lawrence Rosenberg

AMERICAN COOKING GUILD

Boynton Beach, Florida

Dedication
This book is dedicated to my parents for their love. A very special thanks to my friend Dr. Leo De La Cruz (Lee) for his friendship, love and support, to Spiro, for the finest berries in Manhattan. Thanks also to Mark Yackow, Bobby & Herman Bosboom, Patty Kielawa, my friends at the Fairfield Exchange, James Macagna, Nick Lamonica, Nick Katos, Patricia Mack, Debbie Lynch, Lou Weiss, and Alan Applebaum. A very special thank you to Laszlo for a fabulous cover. Thanks to Karen and Dan for working with me around all the obstacles and for believing.

Acknowledgments
—Edited by Polly Clingerman
—Recipe testing by Polly Clingerman and Joan Patten
—Cover Design and Layout by Pearl & Associates, Inc.
—Cover Photo by Laszlo Studios, Inc.
—Food styling by Laszlo and Rosenberg
—Design and typesetting by Clara Graves Graphic Design

Revised Edition 1997
Copyright © 1985 by Lawrence Rosenberg
All rights reserved.
Printed in U.S.A.
ISBN 0-942320-48-4

More...Quick Recipes for Creative Cooking!
The American Cooking Guild's *Collector's Series* includes over 30 popular cooking topics such as Barbeque, Breakfast & Brunches, Chicken, Cookies, Hors d' Oeuvres, Seafood, Tea, Coffee, Pasta, Pizza, Salads, Italian and many more. Each book contains more than 50 selected recipes. For a catalog of these and many other full sized cookbooks, send $1 to the address below and a coupon will be included for $1 off your first order.

Cookbooks Make Great Premiums!
The American Cooking Guild has been the premier publisher of private label and custom cookbooks since 1981. Retailers, manufacturers, and food companies have all chosen The American Cooking Guild to publish their premium and promotional cookbooks. For further information on our special market programs, please contact the address below.

The American Cooking Guild
3600-K South Congress Avenue
Boynton Beach, FL 33426

Contents

Sauces, Soups & Shakes

Introduction

I have always been addicted to berries. My favorites are raspberries, blackberries, and golden raspberries.

I remember as a child going to the Catskill Mountains each summer with my parents for vacation. I picked blueberries, blackberries, and raspberries with Dad and watched Mom turn them into pies and preserves. Having picked berries, I learned to appreciate their cost. The work is backbreaking and extremely tiresome!

Quite versatile, berries add wonderful depth and flavor not only to desserts, but also to salads, soups, and main dishes. In these chapters you will find many interesting and exciting sauces, conserves, syrups, purees, jams, jellies, soups, salads, entrees, and desserts. I have spent nearly three years developing these recipes, carefully testing and tasting each one.

Try my Blueberry Syrup or Raspberry Sauce on your favorite pancakes or French toast, or perhaps over ice cream. Serve the Jazzed-Up Cranberry Relish as an accompaniment for chicken, turkey, pork, duck, fresh ham, or Cornish hens.

If you have a sweet tooth, don't miss the scrumptious pies, cakes, and brownies. On hot summer days, a cold berry soup is delicious and refreshing as a first course or a light dessert.

In closing, I hope that you, and your family and friends enjoy these recipes and I wish you only the berry best!

About Berries

Here are descriptions of the most common types of berries:

Blackberry

A shiny black fruit that decorates country hedgerows in early autumn, the blackberry is also called a bramble because it grows on thorny bushes or brambles. Largest of the wild berries, blackberries are widely cultivated in the United States and are available, depending on the region, from May through August. Purplish-black in color, they range from ½ to 1 inch long when mature. Look for plump, deep-colored berries sans hull. (If the hulls are still attached, the berries are immature and were picked too early; the flavor will be tart.) Eaten fresh, blackberries make a tart, refreshing dessert, but the majority are preserved as jam or jelly. Fresh blackberries are best used immediately, but they may be refrigerated.

Black Currant

Usually sold stripped from their stalks, these summer fruits are always served cooked, as a filling for pies and desserts. They also make excellent preserves, and form the basis of the famous French liqueur cassis.

Blueberry

Round and smooth-skinned, these blue-black berries are juicy and sweet. There are two main types of blueberries. The high bush variety can grow up to 15 feet in height; the hardy low bush variety is only about a foot high and thrives in Canada and the northern United States. Originally wild, this somewhat tart fruit is now grown commercially. Cultivated blueberries comprise the majority of those that reach the market, and the season can span from the end of May to early October. Large New Zealand blueberries are in markets in the winter at a premium price. Blueberries are delicious eaten raw with sugar and cream, stewed, made into soups, preserves, and jams, or used in pies. Choose berries that are firm. Do not wash them until ready to use.

Cranberry

These shiny scarlet berries are grown in big sand bogs on low trailing vines. Cranberries grow wild in northern Europe and in northern regions of America. They are extensively cultivated in Massachusetts, Wisconsin, Washington, and Oregon. Harvested September through early November, the peak market period is around Thanksgiving and Christmas. They are most often used for cranberry sauce, the traditional accompaniment to turkey. They can also be used in cakes, breads, pies, ices, liqueurs, and jellies. Available fresh in the winter, they can be bought frozen or dried. Any cranberries that are discolored or shriveled should be discarded. Cranberries can be refrigerated, tightly wrapped, for at least two months or frozen up to a year.

Gooseberry

Known and enjoyed in Europe since the Middle Ages, gooseberries are summer fruits with a very short season, lasting only a few weeks. Sweet varieties are delicious eaten raw; tart ones make an excellent preserve and can be used in many desserts.

Loganberry

A cross between the raspberry and the blackberry, the loganberry, invented in California by a Scotsman, embodies the best of each.

Raspberry

There are three main varieties of raspberries: black, golden, and red. Intensely flavored, the raspberry is composed of many individual sections of fruit all surrounding a central core. Raspberries often appear as two crops, one in summer and another in autumn. The latter are often smaller, but juicier. Raspberries are fragile and must be handled gently. They can be eaten on their own, with just cream and sugar, or made into soups, purees, fine preserves, sorbets, and other desserts.

Red Currant

Rather tart for eating raw, these summer fruits have many other uses. They make a sparkling jelly; are excellent with roast lamb, poultry, and game; and become a delicious summer salad when tossed with grated raw vegetables. Dipped by the bunch in lightly beaten egg white, then "frosted" with fine sugar, they make a simple but dramatic decoration. They are also used in jams.

Strawberry

Red, juicy, and cone shaped, the strawberry is a member of the rose family and has grown wild for centuries in both the Americas and Europe. The most common American variety is the result of several centuries of crossbreeding of the wild Virginia strawberry. Do not wash the berries until ready to use.

Wild Strawberry

A variety of strawberry, smaller and more aromatic than the ordinary cultivated variety, it is usually eaten fresh and does not need to be hulled before eating. The tiny exquisitely sweet wild strawberries of France are known as fraises des bois or "strawberries of the woods" and are considered by many to be the queen of strawberries.

Fresh Berry Availability

Most berries are available during summer months, mainly June and July. Strawberries are available almost year-round, but the best supply is from April to July. Cranberries are in your market beginning in September through the holidays:

Blackberries	May to September
Blueberries	May to October
Cranberries	September to February
Gooseberries	June to August
Raspberries	May to September
Strawberries	May to December

Best Time to Buy Fresh Berries

Here's when berries will be at peak flavor and ripeness:

Blackberries	July
Blueberries	July to August
Cranberries	October to January
Gooseberries	July to August
Raspberries	July
Strawberries	April to July

Selection and Storage of Berries

Look for plump, fresh, uniformly colored fruit, free of stems and leaves. Avoid fruit that is moldy, crushed, or bruised or that has leaked moisture through the carton.

To Store: Refrigerate and use within one or two days.

To Prepare: Wash berries under cold running water, discarding any crushed or bruised fruit. Drain well in a colander or strainer. Remove stems or hulls from strawberries. Using kitchen scissors, cut top and tail from gooseberries.

Canning Berries

Canning is a wonderful way to preserve excess berries for later use in baking. Consult a book on canning and preserving for detailed instructions:

All Berries (Cold Pack)

Use any variety of edible berry, except strawberries. Wash firm, ripe, freshly picked berries. Drain well. Pack into hot jars as tightly as possible without crushing. Cover with hot syrup made of two parts sugar and one part berry juice or water. Process 20 minutes in hot-water bath, then complete sealing.

All Berries (Hot Pack)

Use any variety of edible berry, except strawberries. Wash and measure firm, ripe, freshly picked berries. Drain. Use ½ to ¾ cup sugar to 1 quart berries. Place layers of sugar and berries in a broad, shallow pan. Let stand 2 hours, then cook by simmering until the sugar is dissolved and the berries are heated through. Pack into hot jars. Process in a hot-water bath, then complete sealing.

All Berries (Open Kettle)

Wash berries. Add from ½ to ¾ cup sugar for each quart of berries. Boil 5 minutes. Pour boiling hot mixture into hot jars and seal at once.

Blueberries and Huckleberries

These berries may be canned by the methods above, but I recommend the following if the berries are to be baked in muffins: Place 2-3 quarts clean berries in a square of cheesecloth. Hold the cloth by the corners and dip into boiling water until spots appear. Then dip quickly in and out of cold water. Pack into hot jars. (Do not add sugar or liquid.) Process 20 minutes in a hot-water bath, then complete sealing.

Elderberries

Wash, stem, and pack into hot jars. Cover with syrup made with 1 part sugar to 3 parts water. Process 20 minutes in hot-water bath; then complete sealing. Elderberries lack acidity, therefore a small amount of lemon juice or vinegar is generally added to them when making pies.

Gooseberries

Gooseberries may be canned by recipes given for all berries, or they can be canned in a heavier syrup (3 parts sugar to 1 part water). Snip heads and tails from clean berries. Pack into hot jars. Cover with boiling syrup. Process 20 minutes in hot-water bath; then complete sealing.

Gooseberries (Open Kettle)

Wash, drain, stem, and measure gooseberries. Add from ½ to 1 cup sugar for each quart berries. Heat slowly to boiling, then boil rapidly for 15 minutes. Pour into hot jars; seal at once.

Red Raspberries

Red raspberries may be canned by any berry recipe, but they keep color and shape better if directions are followed for canning strawberries.

Strawberries

Use berries having small cells and deep red color throughout. Wash, hull, drain, and measure berries. Work with batches of 2 or 3 quarts. Use from 1 to 2 cups sugar to each quart of berries. Place sugar and berries in alternate layers in a broad-bottomed pan. Let stand 2 hours. Simmer 5 minutes without stirring. Cover and let stand overnight. Pack cold berries into hot jars. Process 10 minutes in hot-water bath; then complete sealing.

Cranberry Sauce

Wash and stem cranberries. Boil 4 cups sugar with 4 cups water for 5 minutes. Add 2 quarts berries; boil without stirring until skins burst. Pour into hot jars; seal at once.

Main Dishes

The Berry Book

Pork Ribs with Raspberry Barbecue Sauce

There are a lot of ingredients in the barbecue sauce, but don't let that put you off. It makes the best ribs you'll ever eat and it's delicious with most everything else you normally barbecue. Store extra sauce in the refrigerator—it will last for several months.

1 cup raspberries, washed and cleaned
¼ cup freshly squeezed lemon juice
¼ cup sweet pickle juice
½ cup brown sugar
4 teaspoons A-1® sauce
2½ Tablespoons Worcestershire sauce
1 Tablespoon red wine vinegar
1 small clove garlic, finely chopped
1 Tablespoon finely chopped white onion
¼ cup Pickapeppa® sauce
2 cups bottled barbecue sauce
6 pounds baby back pork ribs
salt and pepper, to taste

To make barbecue sauce: combine all ingredients except ribs in a heavy-bottomed, 2-quart saucepan. Bring to a boil. Turn heat to low and simmer very gently, uncovered, for 1½ hours. Stir every 15 minutes.

Preheat the oven to 450°. Place ribs in a deep pan, cover with cold water, and season with salt and pepper. Bake uncovered for 2 hours. Cool in the liquid. Drain the ribs, dry them thoroughly, and coat with the sauce.

Grill over charcoal or broil, turning ribs frequently and brushing with more sauce until they are well glazed, 6 to 8 minutes per side.

Yield: 6 servings plus extra sauce

Note: Pickapeppa® sauce is sold at most supermarkets in the section where you find Worcestershire and A-l® sauce.

Raspberry Chicken Sauté

This is strictly company fare. The sauce is velvety and the creme de cassis gives the sauce a rich, deep flavor.

2 Tablespoons olive oil, divided
4 chicken breast halves, skinned and boned (2 pounds total)
salt and pepper, to taste
3 Tablespoons chopped shallots
¼ cup raspberry vinegar
¼ cup chicken stock
¼ cup whipping cream or creme fraiche
2 Tablespoons creme de cassis
16 fresh raspberries
fresh chervil or mint leaves for garnish

Heat 1 Tablespoon olive oil in a 10-inch skillet. When hot, sauté chicken breasts for 4 to 6 minutes per side, or until golden brown. Season with salt and pepper. Remove chicken from the pan and keep warm.

Add remaining olive oil to the pan and sauté the shallots over medium heat until translucent, about 2 minutes. Add the raspberry vinegar and cook over high heat until it has almost disappeared.

Add the chicken stock and continue cooking until it reduces by two-thirds. Add the cream. Reduce again until the sauce is thick. Stir in the creme de cassis. Strain the sauce and return it to the skillet.

Return the chicken breasts to the skillet and warm in the sauce for about 5 minutes. Remove chicken to serving plates. Drop the raspberries into the sauce and let them heat. Pour sauce and berries over chicken, dividing evenly. Garnish with fresh chervil or mint leaves.

Yield: 4 servings

Raspberry Barbecued Salmon

The zesty, raspberry-perfumed barbecue sauce adds body and a hint of sweetness to the delicate salmon. The sugar in the sauce gives the fish wonderfully crisp edges. A delicious combination.

10 oz fettuccine
4 salmon steaks, about 1-inch thick
⅓ cup Raspberry Barbecue Sauce, see page 12
1 cup chopped watercress
2 Tablespoons parsley, minced
3 green onions, minced
2 Tablespoons butter
1 teaspoon Worcestershire sauce
16 fresh raspberries and 4 lemon wedges, as garnish

Preheat the broiler. Cook the pasta as directed on the package.

Broil the salmon 4 inches from the heat until it begins to turn white and opaque, about 1 to 2 minutes. Turn and broil the second side, just until the meat turns opaque.

Turn salmon steaks and spread with Raspberry Barbecue Sauce, using 2 to 3 teaspoons per fish steak. Broil until the surface is bubbling and glazed, about 4 minutes. Turn, spread with sauce, and broil another 4 minutes or until glazed.

When the pasta is done, drain well and mix it with watercress, parsley, green onions, butter, and Worcestershire sauce. Make a bed of pasta on each plate. Top with the hot salmon. Garnish each plate with a few berries and a lemon wedge.

Yield: 4 servings

The sweetest raspberries will be a deep wine color, plump and unblemished. Avoid berries whose containers are stained by leaking juice. They will be overripe or damaged and will taste musty. Pale pink berries aren't ripe and will be too tart.

Cranberry Glazed Roast Pork

Tart cranberries are a perfect foil for the moist, rich flavor of pork. This is a wonderful autumn or winter party dish.

1 boneless pork loin roast, about 2 pounds
salt and freshly ground black pepper, to taste
1 teaspoon cornstarch
⅛ teaspoon cinnamon
½ teaspoon grated orange peel
1 Tablespoon orange juice
1 Tablespoon dry madiera
1 cup canned whole berry cranberry sauce

Preheat oven to 325°.

Place the roast in a shallow baking dish, fat side up. Salt and pepper the roast, going heavy on the pepper. Roast uncovered in a 325° oven for 45 minutes.

While the meat is roasting, in a small saucepan combine cornstarch, cinnamon, pinch of salt, grated orange peel, orange juice, madiera, and cranberry sauce. Cook and stir over medium heat until thickened.

When the meat has roasted for 45 minutes, spoon ¼ cup of glaze over it. Roast for 30 to 45 minutes longer, or until the internal temperature registers 160° to 170°. (Remember, today's pork is quite lean and should be served juicy.)

Remove from the oven and let the meat stand for 10 minutes before slicing. Serve with the delicious pan juices and the remaining cranberry-madiera sauce.

Yield: 8 servings

Because of their high acid content, cranberries keep indefinitely. For this reason they were the first North American fruits to be shipped commercially to Europe. In 18th century London they fetched a fancy price.

Steak with Blueberry Cream

Blueberry sauce on steak may sound funny, but this recipe from the French Auvergne is delicious. The slightly tart fruit adds a haunting sweetness to the sauce.

4 8-ounce boneless beef steaks, 1-inch thick
1 Tablespoon oil
1 Tablespoon butter
salt and pepper, to taste
1 shallot, minced
6 Tablespoons dry white wine
½ cup whipping cream
1-2 Tablespoons blueberry preserves

In a large skillet over medium-high heat, sauté the steaks in oil and butter for 3 to 5 minutes per side or until done to taste. Salt and pepper steaks halfway through cooking. Remove steaks from pan and keep warm in an oven on low heat.

Pour all but 1 Tablespoon of fat from the skillet. Add shallot. Sauté 1 minute. Add white wine and scrape up all the browned juices. Simmer until wine is reduced and syrupy.

Add cream and cook over medium-high heat, stirring often. As soon as the cream starts to boil add 1 tablespoon of preserves. Stir until preserves are melted. Taste and add salt if needed. Pour any accumulated juices from the meat into the sauce. Taste again. If sauce is too thin, thicken with a bit more cream and preserves. To serve, spoon sauce over each steak.

Yield: 4 servings

Here's how to freeze berries. Line a baking sheet with wax paper and spread the washed berries on it in a single layer. Freeze solid. Store the frozen berries in an airtight container.

Raspberry-Mandarin Boned Chicken over Spinach

This whole, boned chicken with its unusual, slightly oriental sauce is a breeze to carve. You just slice it into rounds and serve it on a bed of lightly braised spinach.

> 1 whole boned chicken or 3 pounds boneless skinless chicken breasts
> 3 Tablespoons Chinese duck sauce (see note)
> 1 cup fresh raspberries
> 1 10-ounce can mandarin oranges with their juice
> 2 teaspoons raspberry vinegar
> salt and pepper to taste
> 1 bag fresh spinach
> 3 cloves garlic, minced
> 2 Tablespoons olive oil
> 3 Tablespoons warm water
> 1 Tablespoon freshly grated Parmesan cheese
> fresh mint leaves for garnish

Preheat the oven to 350°.

Bone a whole chicken or have the butcher bone a chicken for you. Roll the boned chicken up, starting from the longer side, so that it is shaped like a football. Tie it with butcher's twine. If using chicken breasts, place them in a rectangle shape on waxed paper. Roll up from the long side to create football shape. Carefully tie with butcher's twine.

Place the rolled chicken in a shallow ovenproof baking dish. Baste the chicken with duck sauce. Set aside three raspberries for each diner and place the rest along with the orange segments (save the juice) on top of the chicken. Pour reserved mandarin orange juice and the raspberry vinegar around the chicken. Sprinkle with salt and plenty of freshly ground black pepper.

Bake the chicken at 350° for 45 minutes, or until the internal temperature registers 175°. Remove from the oven and let it sit for 10 minutes. Slice ¼-inch thick and keep warm.

Meanwhile, rinse the spinach under cold water to remove any sand. When the chicken is nearly done, sauté garlic quickly in the olive oil in a medium skillet over medium heat (don't let it brown). Add the water and spinach. Sauté, flipping and stirring, for 3 to 4 minutes. Sprinkle with fresh Parmesan cheese.

To serve, place a portion of spinach in the center of each of six serving plates. Top the spinach with three to four slices of chicken and spoon a little of the savory chicken pan juices over each serving. Garnish each plate with fresh mint and the reserved raspberries.

Yield: 4 to 6 servings

Note: Duck sauce is available in the oriental section of many supermarkets or in stores that carry oriental groceries.

Stuffed Pork Loin with Berries and Prunes

This roast cooks up deliciously moist. The caramelized pan juices make a superb sauce. Save this recipe for your most important dinner parties.

> ½ cup pitted prunes
> 1 cup mixed berries
> 1 boneless pork loin roast, about 2 pounds
> 2 cloves garlic, sliced into thin slivers
> salt and black pepper, to taste
> ¼ cup softened butter
> ½ teaspoon oregano
> ½ cup medium-dry sherry
> 1 Tablespoon brown sugar
> 1 Tablespoon raspberry jelly

Preheat oven to 350°.

Coarsely chop the prunes and combine with the berries. Using a sharp knife, make a lengthwise cut down the center of the pork, cutting ¾ of the way through. Open the boned meat out like you would a book.

Lay the fruit mixture down the center, fold the meat over and tie it into a compact roll with kitchen string: first tie it at 1½-inch intervals crosswise, then run the string around it end to end and tie again. Make slits in the roast and push in the garlic slivers. Salt and pepper generously. Place the meat in a shallow roasting pan and rub the softened butter all over the surface. Sprinkle oregano on top.

In a small bowl, combine the sherry, brown sugar, and raspberry jelly. Pour over the roast. Bake uncovered, basting often, for 40 to 50 minutes, or to an internal temperature of 165° to 170°.

Let meat stand for 10 minutes. Then slice and place on a heated platter. Ladle some of the pan juices down the center of the meat slices and serve the rest separately.

Yield: 6 to 8 servings

Cranberry Lemon Chicken

Don't let the amount of garlic in the recipe put you off. The long cooking turns it sweet and mild and blends it with the sweet dried cranberries and the tart lemon juice into a heavenly sauce.

> 1 chicken, about 3 pounds, cut into serving pieces
> salt and pepper, to taste
> 2 Tablespoons finely minced garlic
> 2 Tablespoons olive oil
> ½ cup dried cranberries
> 3 Tablespoons lemon juice
> ¼ cup chicken broth
> 1 Tablespoon finely grated lemon rind
> 1 teaspoon dried oregano
> ½ teaspoon dried thyme

Preheat oven to 400°.

Salt and pepper the chicken pieces and rub them all over with the garlic. Place the olive oil and dried cranberries in a shallow baking dish that will hold the chicken pieces in one layer, distributing the berries evenly. Place the chicken in the dish skin side down.

Drizzle the lemon juice over the chicken. Pour the chicken broth around the edges. Sprinkle with lemon peel, oregano, and thyme.

Bake the chicken at 400° for 30 minutes. Turn the chicken pieces and bake for another 30 minutes or until done.

Yield: 4 servings

Salads & Side Dishes

The Berry Book

Tossed Salad with Strawberries and Cheese

Serve as a first course or as a main dish for a light meal. Crunchy spinach and radicchio are sprinkled with nippy cheese cubes, creamy bananas, and tangy strawberries in a vinaigrette.

Salad:

3 ounces of fresh spinach, washed and dried
8 radicchio leaves
2 green onions, shredded into long strips
2 small bananas, sliced
1 Tablespoon lemon juice
8 large strawberries, hulled
2 ounces blue cheese, crumbled
2 ounces Brie cheese, cut in cubes
2 ounces Edam or Muenster cheese, cut in cubes

Dressing:

1 clove garlic, crushed
1 Tablespoon white wine vinegar
3 Tablespoons olive oil
salt and pepper, to taste

To make the salad: in a bowl combine the spinach, radicchio, and green onions. Sprinkle the sliced bananas with lemon juice and add them to the salad leaves along with the strawberries. Arrange the salad in a serving dish or on four individual plates. Distribute the cheeses over the top.

To make the dressing: combine the garlic, vinegar, olive oil, salt, and pepper in a screw-top jar and shake until blended. Pour over the salad.

Yield: 4 servings

Strawberry Romaine Salad with Sesame Dressing

Crunchy sautéed pecans add interest to this one. The sesame dressing made with raspberry vinegar is slightly sweet and wonderful.

Dressing:

1 Tablespoon sugar
¼ teaspoon salt
¼ teaspoon dry mustard
1 Tablespoon plus 1 teaspoon raspberry vinegar
1 green onion, chopped
¼ cup salad oil
3 teaspoons sesame seeds

Salad:

2 teaspoons butter
¼ cup pecans
1 head of romaine, washed and torn into bite-sized pieces
1 cup celery, sliced on the diagonal
1 pint strawberries, washed, dried, and hulled

To make the dressing: in a small bowl combine sugar, salt, mustard, vinegar, and green onions. Whisk in oil, then sesame seeds. Refrigerate until needed.

To make the salad: place the butter in a small skillet over medium-low heat. When it is hot, add the pecans and sauté gently until browned on both sides. (Watch carefully—once browning starts it goes very quickly.) Drain pecans on paper towels.

In a bowl combine romaine and celery. Cut the strawberries in half and add to the salad. Top with the pecans. Add dressing and toss gently to coat well.

Yield: 4 servings

Green Salad with Avocado and Raspberry Vinaigrette

Creamy avocado, fragrant raspberries, and crisp greens combined with a sweet-tart raspberry vinaigrette make this salad a real treat.

Dressing:

1 Tablespoon chopped fresh parsley
2 Tablespoons olive oil
1 Tablespoon raspberry vinegar
2 teaspoons honey
1¼ teaspoons Dijon mustard
salt and freshly ground pepper, to taste

Salad:

2½ cups romaine lettuce, washed and torn into bite-sized pieces
2½ cups Boston lettuce, washed and torn into bite-sized pieces
2 cups watercress
½ cup peeled, seeded, and sliced cucumber
1 avocado, peeled and sliced
4 big parsley sprigs, minced
1 cup fresh raspberries

In a small bowl whisk together the chopped parsley, olive oil, vinegar, honey, mustard, and salt and pepper until blended.

In a large salad bowl combine the romaine, Boston lettuce, watercress, cucumber, avocado, and parsley sprigs. Pour dressing over and toss gently. Top with raspberries.

Yield: 4 servings

Green Salad with Oranges and Cooked Raspberry Dressing

Pale Boston lettuce, dark green watercress, and bright oranges with a ruby colored dressing combine to form a very pretty salad. The fruit flavors add zip to the crisp greens.

Dressing:

¾ cup fresh raspberries
3 Tablespoons sugar
1 Tablespoon cornstarch
1 cup water
2 Tablespoons fresh lemon juice
2 Tablespoons raspberry vinegar
1 Tablespoon dry white wine
3 Tablespoons fruity olive oil

Salad:

5 cups Boston lettuce, washed and torn into bite-sized pieces
3 cups watercress
2 cups fresh orange sections
2 Tablespoons sliced almonds, as garnish

In a heavy saucepan combine the raspberries, sugar, cornstarch, water, lemon juice, raspberry vinegar, white wine, and olive oil. Bring to a boil over medium heat stirring constantly. Boil and stir for 1 minute. Remove from heat and strain. Discard seeds and pulp. Pour the dressing into a covered container and chill.

To make the salad: arrange lettuce on six salad plates. Arrange watercress and orange slices on the leaves. Spoon dressing over the salad and sprinkle with almonds.

Yield: 6 servings

Crispy Chicken and Spinach Salad

This is a salad of contrasts. Hot, spicy, crisp-coated chicken lies on a bed of cold salad beneath a raspberry vinegar dressing. Quick, light, and easy, this is an excellent dish for a light supper. Serve it with French or Italian bread or hot corn muffins.

Chicken:

8 chicken tenders or 2 whole chicken breasts
4 teaspoons creole seasoning (see note)
½ teaspoon black pepper, or to taste
½ teaspoon white pepper, or to taste
½ cup buttermilk
½ cup all-purpose flour
1 Tablespoon vegetable oil
1 Tablespoon butter

Dressing:

½ cup olive oil
3 Tablespoons raspberry vinegar
1 teaspoon black pepper
½ teaspoon dried oregano
½ teaspoon salt
1 clove garlic, minced
¼ cup green onions, sliced

Salad:

2 large carrots, julienned
1 pound spinach, washed and dried
1 avocado, peeled and cut in wedges
1 cup fresh raspberries

If you use whole breasts, cut each lengthwise into four strips. Rub creole seasoning, black pepper, and white pepper into the chicken tenders. Dip chicken in buttermilk, then flour to coat completely. Shake off any excess flour. Place coated chicken on a rack to dry while you make the salad.

In a small bowl whisk together olive oil, vinegar, black pepper, oregano, salt, garlic, and green onions.

Drop julienned carrots into boiling water and boil uncovered for 3 minutes. Drain immediately and rinse with cold water to stop the cooking. Blot with paper towels. In a large bowl, combine the spinach, avocado, and carrots. Toss with the dressing. Set aside.

In a large skillet heat 1 tablespoon oil and 1 tablespoon butter. Fry chicken until nicely browned on both sides, about 4 to 6 minutes. Drain cooked chicken on paper towels.

Divide the salad onto four plates. Top each with two pieces of chicken. Scatter raspberries over the salads.

Yield: 4 servings

Note: If you can't find creole seasoning, make your own. Combine 1 teaspoon salt, ¾ teaspoon oregano, ½ teaspoon basil, ¾ teaspoon thyme, ½ teaspoon onion powder, ½ teaspoon garlic powder, and ½ teaspoon paprika. Mix well.

Raspberry Marinated Carrots

This cross between vegetable and salad is a deliciously different way to serve carrots. Hazelnut oil has a distinctive flavor that is not to everyone's taste. If you don't enjoy its slightly musky flavor, olive oil is an excellent substitute.

> 1½ lbs carrots
> ⅓ cup raspberry vinegar
> ½ cup hazelnut or olive oil
> salt and pepper, to taste

Wash and scrape the carrots and cut them in julienne (matchstick-sized pieces). Drop carrots in boiling water and boil them for 3 minutes. Drain immediately and plunge into cold water to stop the cooking. Drain well, then pat dry on a triple layer of paper towels.

Place carrots in a salad bowl, add the raspberry vinegar and toss to coat well. Chill in the refrigerator for at least 1 hour (overnight is better). Before serving, add the oil, and salt and pepper to taste. Serve chilled.

Yield: 6 to 8 servings

Strawberry Antipasto

Here's a different first course, perfect for summer. It combines the sweetness of fruits, the suaveness of avocado and pungent garlic, Parmesan, and prosciutto. It also makes an excellent light summer luncheon.

Dressing:

¼ cup fruity olive oil

2 Tablespoons water

3 Tablespoons red wine vinegar

2 Tablespoons chopped fresh mint (or 2 teaspoons dried, crushed)

2 teaspoons Dijon mustard

2½ teaspoons honey

1 clove garlic, minced

¼ teaspoon salt

¼ teaspoon pepper

Salad:

1 small Persian melon or cantaloupe, rind and seeds removed

2 avocados, peeled and pitted

2 nectarines or peaches, peeled and pitted

1 pint fresh strawberries, rinsed and dried

1 orange, peeled, sliced in ½-inch rounds and halved

4 ounces Parmesan cheese, shaved into very thin slices

4 ounces prosciutto ham, in very thin slices

In a small bowl whisk together the olive oil, water, vinegar, mint, mustard, honey, garlic, and salt and pepper.

Cut melon, avocados and nectarines into wedges. Arrange fruits attractively on a large platter or on six individual plates. Add slices of Parmesan and prosciutto. Spoon the vinaigrette over all. Serve at room temperature.

Yield: 6 servings

Potato Pancakes
with Strawberry Applesauce

Serve these pancakes as a vegetable with roast meat or poultry or serve three to a person for an unusual first course. These taste best if they don't wait so fry them at the last minute.

Strawberry Applesauce:

2 medium golden delicious apples

3 cups fresh or frozen unsweetened strawberries

1 teaspoon grated lemon rind

½ cup sugar

½ cup water

Potato Pancakes:

2 pounds baking potatoes

2 eggs, slightly beaten

1 cup diced red onions

½ cup chives, minced

salt and pepper, to taste

½ cup vegetable oil, for frying the pancakes

Sour cream, as garnish

To make the applesauce: peel and core the apples and cut in thin slices. Wash and stem the strawberries. Puree the strawberries in the food processor. Place apples, berry puree, lemon rind, sugar, and water in a saucepan. Cook uncovered over medium heat until apples are soft, about 25 minutes. **Yield:** about 3 cups

To make the pancakes: peel and julienne the potatoes. Place potatoes in paper towels, and squeeze out the excess water. Place potatoes in a bowl. Add eggs, onions, chives, and salt and pepper to taste.

Preheat oven to 200°. Form potatoes into 3-inch cakes (about 2 rounded tablespoons). Heat 2 to 3 tablespoons vegetable oil in a large skillet over medium-high heat. Sauté 2 or 3 pancakes at a time until cakes have a golden brown crust, about 5 minutes per side. Drain on paper towels and keep warm in the oven until all are fried.

Yield: 16 pancakes

Navy Beans in Raspberry Vinaigrette

This combination of bland beans in a peppy sauce of minced onion, raspberry vinegar and fresh basil is irresistible. It is tangy, fragrant, and subtle, but still down to earth. Serve it as a vegetable or salad.

1 can (15 ounces) cannellini or navy beans
1 cup finely chopped mild onion
1 teaspoon garlic, finely minced
2 Tablespoons parsley, finely chopped
2 Tablespoons fresh basil, chopped
1 Tablespoon fresh mint, chopped
¾ teaspoon salt
freshly ground black pepper
1 teaspoon sugar
¼ cup raspberry vinegar
¼ cup fruity olive oil
½ cup fresh raspberries, rinsed and drained

Drain beans, discarding the liquid, and rinse with cold water.

In a 1-quart saucepan mix onion, garlic, parsley, basil, mint, salt, pepper, sugar, vinegar, and olive oil. Mix well. Add the beans and mix lightly so as not to crush or break them. Heat over very low heat until warmed through. Serve warm, or at room temperature. Just before serving, gently fold in the raspberries.

Yield: 4 servings

Acorn Squash Stuffed with Blueberries

This is a very pretty dish of deep orange squash heaped with glistening purple berries.

2 acorn squash, about 1 pound 4 ounces each
salt and pepper, to taste
cinnamon, to taste
1 small apple, peeled, cored, and chopped medium fine
8 teaspoons brown sugar
4 teaspoons butter
1½ cups fresh or frozen unsweetened blueberries

Preheat oven to 375°.

Cut squash in half lengthwise. Scoop out and discard the seeds and strings. Cut a little slice from the bottom of each so it sits level. Sprinkle cut sides with salt, pepper, and cinnamon.

Divide the chopped apple evenly among the squash halves. Sprinkle each with 2 teaspoons brown sugar and a light shake of cinnamon. Dot each with 1 teaspoon butter. Place the squash halves in a shallow ovenproof dish and pour ½ inch of water around them. Cover the dish with foil. Bake at 375° for 40 to 45 minutes or until squash is tender when pierced with a fork. Remove the foil.

Put an equal portion of berries in each squash half. Return the dish to the oven and bake uncovered at 375° for 10 minutes more.

Yield: 4 servings

To freeze fresh strawberries with sugar, wash and hull them. Gently stir 2 cups of small whole berries in a bowl with ⅓ cup sugar. Mix well. Pack in a rigid container.

Cranberry-Chive Tartlettes

These are flat, buttery pastries topped with a sweet-tart dome of cranberries. They make a superb accompaniment to roast poultry or meat. Try them with Thanksgiving turkey, a crisp-crusted roast pork loin, or game hens.

Pastry:

1 cup all-purpose flour
¼ teaspoon salt
⅛ teaspoon freshly ground black pepper
1 teaspoon fresh chives
¼ cup butter
2 Tablespoons solid vegetable shortening
2-3 Tablespoons ice water

Compote:

3 cups water
3 cups granulated sugar
3 cups fresh or frozen cranberries
1 teaspoon cinnamon

In a medium bowl combine flour, salt, pepper, and chives. Cut in the butter and vegetable shortening until the mixture resembles coarse meal. Stir in the ice water a tablespoon at a time. You may need a little more than 2 tablespoons. Use only enough to allow the mixture to pull into a ball. Wrap in plastic wrap and chill for 1 hour.

In a 2-quart saucepan bring water and sugar to a boil over high heat. Add the cranberries and cinnamon. Simmer, stirring occasionally, until the compote is thick, about 10 to 15 minutes. Cool.

Preheat the oven to 350°. On a lightly floured board, roll the pastry ¼-inch thick. Stamp out 2-inch rounds and place them on a baking sheet. Mound 1 tablespoon of the cooled cranberry compote on each round. Bake for 28 minutes or until the pastry is golden. Cool on a rack.

You can prepare the tartlettes a day ahead and store them in a cool room in an airtight container. Reheat in a 350° oven until warm, about 5 to 6 minutes.

Yield: about 24 tartlettes

Sweets & Treats

& Treats

The Berry Book

Feather-Light Blueberry Pancakes

These are perhaps the best pancakes you'll ever eat. They are thick, airy, tender, and oozing with fat juicy berries.

1¼ cups all-purpose flour
1 Tablespoon baking powder
1 Tablespoon sugar
½ teaspoon salt
1 egg, beaten
1 cup milk
2 Tablespoons vegetable oil
1 cup fresh or unsweetened frozen blueberries

In a large bowl, whisk together the flour, baking powder, sugar, and salt.

In a small bowl stir together egg, milk, and vegetable oil and add to the dry ingredients, stirring only enough to dampen the flour. Gently fold in the berries.

Heat a griddle or skillet until a drop of water bounces. Grease lightly. For each pancake pour ¼ cup of batter on the griddle. When holes appear in the center and edges start to dry, turn and brown the second side. Keep pancakes warm in a 200° oven while you prepare the rest.

Serve with butter and Blueberry Syrup, page 56.

Yield: 6 to 8 pancakes

Choose blueberries that have a silvery sheen and no tinge of red. They should look clean and very dry—moisture starts the decaying process. Overripe blueberries are soft, watery and often show mold. You can store them for up to 2 weeks in the refrigerator in a rigid container covered with plastic wrap.

Suit-Your-Mood Muffins

This recipe makes three varieties of tender, fruit-filled muffins: cranberry-apple, blueberry, or strawberry-almond. You can substitute any berries in any of the variations.

4 cups all-purpose flour
1½ cups sugar
5 teaspoons baking powder
1 teaspoon salt
3 eggs
2 cups milk
½ cup melted butter
Cranberry, blueberry, or strawberry mixture, see below

Preheat oven to 400°. Grease muffin pan or line with papers.

In a large bowl, whisk together the flour, sugar, baking powder, and salt.

In a small bowl whisk the eggs lightly. Then whisk in the milk and melted butter. With as few strokes as possible, stir the wet ingredients into the flour mixture. Lightly fold in the prepared fruit (see end of recipe for berry mixtures). Spoon batter into each muffin well, about two-thirds full.

Bake at 400° oven for 20 to 25 minutes or until browned and a toothpick inserted in the center comes out clean. Remove the muffins immediately from the pan, and cool on a rack.

Yield: about 24 muffins

Cranberry-Apple: Mix together ½ cup fresh or thawed frozen cranberries, 1 cup chopped apple, ¼ cup sugar, and ½ teaspoon cinnamon.

Blueberry: Mix 1 cup fresh or thawed frozen unsweetened blueberries with 1 teaspoon flour, ⅓ cup sugar, and ¼ teaspoon nutmeg.

Strawberry-Almond: Mix 1 cup sliced strawberries, ¼ cup chopped almonds, ⅓ cup sugar, ¼ teaspoon cinnamon, and ¼ teaspoon nutmeg.

Yield: about 36 muffins

Raspberry Brownies

Raspberry preserves give these rich and chewy brownies an elusive flavor.

2 ounces unsweetened chocolate
½ cup unsalted butter
¼ teaspoon salt
½ teaspoon vanilla extract
1 cup sugar
2 eggs
¾ cup all-purpose flour
¾ cup chopped walnuts
⅓ cup seedless raspberry preserves

Preheat the oven to 350°. Grease and flour an 8-inch baking pan.

Place chocolate and butter in a large heavy saucepan over low heat. Stir frequently until chocolate melts, then remove from heat and cool slightly. Stir in salt, vanilla, and sugar. Add the eggs one at a time, stirring to incorporate each before adding the next. Add the flour and stir to combine. Stir in walnuts.

Pour the batter into the prepared pan. Spoon the raspberry preserves over the batter in small dollops. With a knife, marble preserves through the batter, making an even swirl pattern.

Bake for 35 to 40 minutes or until the edges start to pull from the pan and a toothpick inserted in the middle comes out clean.

Yield: 9 brownies

To remove berry stains from counter tops, pour a little bleach or spray a bleach-type cleanser on the stain, wipe with a paper towel, then rinse with water. To remove berry stains from washable clothing, soak the stain in cool water for several hours to loosen it, then rub with mild soap. If that doesn't do it, *on bleachable fabric only* you can use a weak solution of bleach and water.

Light Cranberry Cheesecake

No one believes this rich cheesecake is actually low in fat.

1½ cups graham cracker crumbs
½ cup Grapenuts® cereal
2 Tablespoons vegetable oil or melted butter
1 16-oz can whole cranberry sauce
3 8-oz packages low-fat cream cheese, softened
¾ cup sugar
¼ cup all-purpose flour
3 eggs
1 cup low-fat dairy sour cream
2 teaspoons vanilla extract
Whole sugared cranberries and mint leaves, as garnish

Preheat oven to 325°.

In a small bowl combine the graham cracker crumbs, Grapenuts® cereal, and oil. Press into the bottom of a 9-inch springform pan. Bake at 325° for 5 to 8 minutes, or until golden. Remove to a rack and let the crust cool 5 minutes. Spread prepared crust with the cranberry sauce.

Reduce the oven temperature to 300°. In a large bowl using an electric mixer, beat the cream cheese, sugar, and flour until smooth. Beat in the eggs one at a time. Then beat in sour cream and vanilla, blending well. Spread the cheese mixture evenly over the cranberry sauce. Bake the cheesecake at 300° until a knife inserted 1½ inches from the edge comes out clean, about 90 minutes. Turn off the oven and leave the cheesecake in the oven with the door ajar for 30 minutes. Remove the cake from the oven and cool on a rack for 1 hour.

Cover the cheesecake with foil or plastic wrap and refrigerate it for at least 4 hours. To serve, run a thin knife blade around the sides of the cake pan, then gently remove the springform and place cake on a serving plate. Garnish with sugared cranberries and mint leaves.

Yield: 10 servings

Sugared Cranberries: Pour a little white corn syrup into a small bowl. Place ½ cup granulated sugar in another bowl. Drop fresh cranberries into the corn syrup, turning to coat. Lift the berries out and drop into the sugar. Toss gently to coat.

Cranberry Nut Loaf

This loaf is moist, not too sweet, and slightly chewy. Serve it slathered with butter for breakfast, with morning coffee, or with a fruit salad for lunch.

2 cups all-purpose flour
⅔ cups sugar
½ teaspoon salt
2 teaspoons baking powder
½ teaspoon baking soda
½ teaspoon cinnamon
⅓ cup shortening
2 eggs
¾ cup milk
1 Tablespoon grated orange rind
⅔ cup fresh or frozen cranberries
⅓ cup walnuts, chopped

Preheat the oven to 350°. Grease and flour a 9-inch loaf pan.

Place ¼ cup of the flour in a small bowl. In a large bowl, whisk together the rest of the flour with the sugar, salt, baking powder, baking soda, and cinnamon. Using a pastry blender or a fork, cut in the shortening.

In a small bowl beat the eggs lightly, then beat in the milk and orange rind. Pour this into the flour mixture, stirring only until moistened. Toss the cranberries and nuts with the reserved flour and fold them into the batter. Pour into the prepared pan.

Bake at 350° for 60 minutes or until well browned and a toothpick inserted near the center comes out clean. Remove from the pan and cool on a rack. Wrap the cooled loaf in foil or plastic wrap. (Loaf will cut easily and taste better if it rests 8 hours before serving.)

Yield: 1 loaf

Blueberry Lattice Pie

A lattice pie is always pretty and is surprisingly easy to do. The crust has ground almonds added and the filling is bright-tasting and satiny.

Almond Pastry:

1 cup all-purpose flour
½ cup cake flour
½ teaspoon salt
2 Tablespoons sugar
6 Tablespoons cold butter, cut into small pieces
⅓ cup vegetable shortening, cut in small pieces
½ cup finely ground almonds
1 egg
1 teaspoon almond extract
¼ cup cold water
1 egg white plus 2 teaspoons sugar, for egg wash

Filling:

3 Tablespoons lemon juice
2 ½ Tablespoons cornstarch
⅔ cup sugar
3 ½ cups fresh or frozen unsweetened blueberries

To make the crust: in a large bowl, whisk together the all-purpose flour, cake flour, salt, and 2 tablespoons of the sugar. Using a pastry blender or fork, cut in the butter and shortening to make even particles the size of oatmeal flakes. Stir in the ground almonds.

In a small bowl beat together the egg, almond extract and water with a fork. Add egg mixture, a tablespoon at a time, to the flour mixture, tossing with a fork to mix. Gather dough into a ball, wrap in plastic, and chill in the refrigerator for at least 1 hour.

Preheat oven to 400°. Cut off about two-thirds of the pastry and refrigerate the remainder. Roll the dough into an 11-inch circle and fit it into a 9-inch pie plate, fluting the edge as desired. With a fork, prick the dough at ½-inch intervals, line the shell with foil and fill with dried beans or rice to weigh down the crust. Bake in the middle level of the oven for 8 minutes. Remove from the oven and lift out the beans and foil.

To make the filling: in a medium bowl combine the lemon juice, cornstarch, and sugar. Gently stir in berries, mixing well. Let stand for 30 minutes.

To make egg wash: in a small bowl beat the egg white with 2 teaspoons sugar. Paint the bottom of the crust with the glaze. Reserve the remainder. Return the crust to the oven and bake for 2 minutes. Remove and cool on a rack. Raise the oven temperature to 425°. Roll out the remaining pastry on a sheet of floured wax paper to form a circle about 10 inches across. With a fluted pastry wheel or a sharp knife cut it into 12 strips about ¾-inch wide.

Pour the filling into the cooled pie shell. Moisten both ends of the longest strip of pastry with water and center it over the berries, pressing the ends down well. Continue with the remaining strips, running half of them in each direction and weaving them over and under in a criss-cross pattern to make a lattice.

Using a fork, beat the egg white-sugar mixture to a froth and brush it gently over the lattice top. Bake the pie on the lower rack of the oven for 15 minutes. Lower the heat to 375° and bake for another 30 minutes or until the top is browned and the filling bubbles. If the edges brown too fast, protect them with strips of foil. Cool pie on a rack before cutting.

Yield: 1 pie

Old-Fashioned Berry Shortcake

This is a wonderful, old-timey shortcake. Of course you can use just one kind of berry, but a mixture of blackberries, blueberries, strawberries, and raspberries is especially dramatic.

Berry Topping:

3 pints fresh mixed berries (about 6 cups)
1 cup sugar

Shortcake:

2 cups all-purpose flour
3 teaspoons baking powder
¾ teaspoon salt
2 Tablespoons sugar
½ cup shortening
¾ cup milk
¼ cup butter, softened
1 cup whipping cream, as topping

Preheat oven to 450°.

Set aside some whole berries for garnish. Slice strawberries. Mix them with the other berries and crush lightly. Stir in 1 cup of sugar. Set aside.

To make the shortcake: whisk together the flour, baking powder, salt, and sugar. Cut in the shortening until it is crumbly. Make a well in the center and pour in the milk. Quickly stir crumbly mixture into the milk until it just holds together. Form a ball and place on a lightly floured board. Knead 15 times. Pat out ½ inch thick. Cut in 3-inch rounds. Spread half the rounds with softened butter. Top with remaining rounds. Place on baking sheet and bake at 450° for 10-15 minutes.

Separate shortcakes and spoon half the berries over the bottom layer. Put on tops and add more berries. Whip the cream and sweeten with sugar, if desired. Spoon whipped cream on top.

Yield: 6 servings

Strawberry Nut Bars

A crisp, butter-cookie base, a layer of strawberry jam topped with chewy, nut-studded meringue. These are especially good during the holiday season.

¾ cup unsalted butter
¾ cup sugar, divided
2 eggs, separated
1 ½ cups all-purpose flour
1 cup chopped walnuts
1 cup top-quality strawberry preserves
½ cup flaked coconut

Preheat the oven to 350°.

In a medium mixing bowl cream the butter with ¼ cup of the sugar until light and fluffy. Beat in the egg yolks one at a time. Stir in the flour until well blended and spread the batter evenly in a 9 x 13-inch pan. Bake at 350° for 15 minutes or until golden. Remove from the oven.

While the cookie layer bakes, beat the egg whites in a small bowl until foamy and doubled in volume. Beat in the remaining ½ cup of sugar and continue to beat until the meringue stands in firm peaks and all sugar is dissolved. Fold in walnuts.

Spread the preserves over the cookie layer. Sprinkle with coconut. Carefully spread the meringue over all. Return the pan to the oven and bake at 350° for 25 minutes or until the top is light gold.

Yield: about 4 dozen cookie bars

The best strawberries are plump, deep red, and smell sweet. Avoid those that are shriveled or have brown leaves—they will be bitter. Berries with pinkish skins aren't ripe and will be sour. Don't buy strawberries in a deep container; the bottom ones will be crushed.

Cranberry Apple Crisp

This is a comforting, grandmother-type pudding. Serve it warm, in pottery bowls if you have them, add a ladle of thick cream or let a scoop of ice cream melt into the fruit and buttery crunch as you eat.

1½ cups graham cracker crumbs
¼ cup sugar
½ cup butter, melted
4 cups tart apples, peeled, cored, and chopped
1½ cups toasted almonds, chopped coarsely
½ cup brown sugar (¾ cup if apples are very tart)
1 Tablespoon dark rum
1 Tablespoon orange liqueur
¼ cup dried cranberries
1 teaspoon cinnamon

Preheat the oven to 350°.

In a small bowl mix graham crackers, sugar, and butter.

In a 2-quart casserole, toss chopped apple with nuts, brown sugar, rum, orange liqueur, cranberries, and cinnamon to mix well. Top with the graham cracker mixture.

Bake at 350° for 35 minutes or until the apples are tender and top is well browned.

Serve warm or cold topped with cream, ice cream, or frozen yogurt.

Yield: 6 to 8 servings

Toasted almonds: place skinned almonds in a single layer on a cookie sheet and bake at 300° oven until pale brown, about 8 to 10 minutes. Watch carefully; once they start to brown they darken fast.

Berried Treasure Cake

This show-stopping cake looks like an opened jewel box with berries as the treasure. It is a lovely dessert for a buffet party because the guests can admire the cake before it is cut. A combination of raspberries, blueberries, and strawberries is beautiful.

Cake:

4 eggs
1 cup superfine sugar
1 teaspoon vanilla extract
⅓ cup water
1 cup self-rising flour (see note)

Filling:

2 Tablespoons seedless raspberry jam
2 cups fresh mixed berries
2 cups whipping cream, whipped
4 teaspoons superfine sugar
¼ cup confectioners' sugar
A few berries with stems, for decoration

Preheat the oven to 375°.

Grease the bottom of a 9 x 13-inch cake pan. Don't grease the sides. Line the bottom of the pan with wax paper. Grease and flour the paper. Set aside.

In a medium bowl beat eggs and sugar until very thick and lemon colored and the batter falls from the raised beaters in a thick ribbon, about 8 to 10 minutes. Beat in the vanilla and water. Sift the flour over the mixture and fold it in very gently so as not to lose the air you've whipped in. Pour the batter into the prepared pan and bake in the center of the oven for 12 to 15 minutes, until it springs back when lightly touched with a finger. Don't overbake. Cool in the pan on a wire rack for 15 to 20 minutes. Run a knife around the edge of the cake and turn it out of the pan and onto a cooling rack.

Place the jam in a small bowl and microwave for 20 seconds to thin it out. Gently toss the berries in the warmed jam. Set aside.

When completely cool, cut cake in half crosswise to make two layers, each 9 x 6½-inches. Place one layer on a serving plate. Whip the cream with 4 teaspoons superfine sugar and spread it on the sides and top of the cake. Position the second cake layer over the first at an angle, so that one edge rests on the plate and it looks like the top of an open jewel box. Carefully frost the sides and top of the angled layer with whipped cream.

Carefully spoon the prepared berries onto the top of the cake, arranging them as if they are jewels in a jewel box. Dust the berries with a very light coating of confectioners' sugar. Garnish the serving plate with a few whole berries with stems. Chill before serving.

Note: If you don't have self-rising flour, substitute 1 cup sifted cake flour plus 1 teaspoon baking powder and ¼ teaspoon salt.

Elegant Raspberry Chocolate Box

A delicate cake filled with cream and berries is surrounded by a thin box of chocolate. Making the chocolate box isn't as hard as it sounds!

Cake:

4 eggs, lightly beaten
½ cup superfine sugar
grated rind of ½ lemon
¾ cup all-purpose flour
¼ cup cornstarch
4 Tablespoons butter, melted

Filling:

2 oranges, juiced
2 Tablespoons orange liqueur
2 cups heavy cream
¼ cup superfine sugar plus 2 Tablespoons, divided
2 lbs fresh raspberries, hulled
1 Tablespoon confectioners' sugar, to garnish cake

Chocolate Box:

6 squares semisweet chocolate
1 Tablespoon solid vegetable shortening

Preheat oven to 350°. Grease and flour an 8-inch square cake pan. Place the eggs, sugar, and lemon rind in the bowl of an electric mixer.

Beat at medium-high speed 8 to 12 minutes, or until the mixture is very thick and pale yellow (it will almost triple in volume).

In a bowl, whisk together the flour and cornstarch. Carefully fold it into the egg mixture. Then fold in the melted butter, taking care not to lose too much air. Pour the batter into the prepared pan and bake at 350° for 30 to 35 minutes or until the cake is golden and springs back from a light touch. Cool in the pan for 10 minutes, then turn out onto a wire rack to cool.

In a small bowl, combine juice of two oranges with 2 tablespoons orange liqueur. Using a knife with a long serrated blade, split the cooled cake horizontally into two layers. Sprinkle both layers with the orange juice-liqueur mixture.

Whip 2 cups heavy cream with ¼ cup superfine sugar until it stands in soft peaks. Measure out and set aside 1 cup whipped cream for the sides of the cake. Set aside a third of the raspberries for decoration. Sweeten remaining berries to taste using 1 to 2 tablespoons superfine sugar. Fold berries into the larger quantity of cream.

Spread the raspberry cream on the bottom cake layer. Top with the second layer and frost top with raspberry cream. Spread a thin layer of the plain, reserved whipped cream around the sides of the cake. Decorate the top of the cake with the whole raspberries. Dust with confectioners' sugar, using a sieve.

To make the chocolate box: Cut four strips of waxed paper, each 8 x 2-inches. Lay the paper strips on a large sheet of aluminum foil. Line a baking sheet with aluminum foil and set it aside. Melt the chocolate and shortening in a small heavy saucepan over very low heat. Pour melted chocolate down the center of each waxed paper strip. Using a metal icing spatula or knife, spread the chocolate thinly over the paper. Gently lift the chocolate-coated strips onto the foil-lined baking sheet. Refrigerate until chocolate begins to set but is still flexible, about 3 minutes. Leaving paper in place, press a chocolate band onto one side of the cake. Repeat with other three bands. (If edges overlap, trim off the excess with scissors.) Leaving paper in place, refrigerate cake until chocolate sets, about 5 minutes. Gently peel off paper. Refrigerate cake until serving time.

Yield: 8 servings

Blueberry Cream Pie

This deep purple, berry-rich pie tastes a bit like cheesecake.

Pie Crust:

1¼ cups all-purpose flour
½ teaspoon salt
¼ cup butter
2 Tablespoons solid vegetable shortening
2-3 Tablespoons ice water

Filling:

1 cup sour cream
1 egg
¾ cup sugar
2 Tablespoons flour
¼ teaspoon salt
½ teaspoon almond extract
1 pint fresh blueberries, rinsed and patted dry (2 cups)

Topping:

6 Tablespoons flour
3 Tablespoons chopped almonds
3 Tablespoons butter

To make the crust: whisk together the flour and salt. Cut in the butter and vegetable shortening until it resembles cornmeal. Starting with 2 tablespoons, add the cold water, tossing the flour-shortening mixture with a fork. Add only enough water to allow the dough to pull together into a ball. Chill 30 minutes. Roll out and line a 9-inch pie plate.

To make the filling: preheat oven to 400°. In a medium bowl beat the sour cream with the egg. Beat in sugar, 2 tablespoons of flour, salt, and almond extract. Fold in the berries. Pour mixture into the pie shell.

Bake at 400° for 25 minutes or until filling puffs and the crust is golden.

To prepare the topping: place the flour, almonds and butter in a small bowl and mix with a fork until crumbly.

Remove baked pie from the oven and sprinkle with topping mixture. Return the pie to the oven and bake for 10 minutes more. Cool the pie on a rack, then refrigerate. Serve chilled.

Yield: 1 pie, about 8 servings

Raspberry Coconut Squares

This buttery crust holds a yummy, chewy raspberry filling.

Crust:

1¼ cups all-purpose flour
¼ teaspoon salt
½ cup cold butter
10-oz jar raspberry jam

Topping:

2 eggs
1 cup sugar
¼ cup melted butter
1 teaspoon vanilla extract
1 can (3½ ounces) flaked coconut

Glaze:

½ cup sifted cocoa
pinch of salt
⅓ cup whipping cream
2½ Tablespoons butter
⅔ cup sugar
½ teaspoon vanilla extract

Preheat oven to 350°. **To make the crust:** whisk together the flour and salt. Cut in the butter until it looks like oatmeal flakes. Press mixture firmly on the bottom of a 9 x 13-inch pan. Bake at 350° for 12 minutes or until golden. Remove from oven and cool slightly. Don't turn off the oven. Spread the raspberry jam on the crust.

To make the bars: in a medium bowl beat together the eggs, sugar, melted butter, and vanilla. Stir in the coconut. Spoon this on the jam layer and spread to cover completely. Return to the oven and bake for 20 minutes or until light gold. Cool.

To make the glaze: in a heavy saucepan combine cocoa, salt, cream, butter, and sugar. Cook and stir over low heat for about 5 minutes or until mixture is thick and smooth. Remove from the heat. When it stops bubbling, stir in the vanilla. Spread cooled cookie bars with the glaze. Chill until glaze sets.

Yield: 20 to 24 bars

Blackberry Harvest Pie

The glistening berry filling is black, sweet, and spicy. This pie says "summer." You can use the Almond Pastry, Standard Pie Crust, or Sweet French Pastry for this pie.

Pastry for 2-crust pie, see pages 38, 46 or 49
4 cups blackberries
¾ cup sugar
3 Tablespoons all-purpose flour
½ teaspoon cinnamon
½ teaspoon nutmeg
pinch salt
2 teaspoons lemon juice (don't use if berries are very tart)
1 Tablespoon butter

Preheat the oven to 400°. Prepare pie crust as recipe directs. Roll out half of the dough and line a 9-inch pie plate.

Mix together berries, sugar, flour, cinnamon, nutmeg, salt, and lemon juice. Pour into the pie shell. Dot with butter. Roll out remaining crust and place on top of the pie. Crimp the edges as desired. Slash the crust in 2 or 3 places so steam can escape.

Bake at 400° for 35 minutes or until crust is golden. Serve warm or at room temperature.

Note: Because of the large amount of sugar in it, this crust browns quickly. If you see this happening, cover the rim of the crust with foil.

Sweet French Pastry Dough

This dough is sweet and buttery, almost like a shortbread cookie.

2 cups all-purpose flour
½ cup sugar
½ teaspoon salt
½ cup butter
2 Tablespoons solid vegetable shortening
2 egg yolks
2-3 Tablespoons ice water

In a large bowl whisk together the flour, sugar, and salt. Cut the butter in small pieces and add it with the vegetable shortening. Work it into the flour quickly using a pastry cutter or your fingers, until each piece of fat is about the size of a raisin. Work quickly before the butter warms and makes the crust greasy.

In a small bowl lightly beat the egg with 2 tablespoons of water. Add it to the flour mixture, tossing with a fork. If the dough doesn't come together, add up to a tablespoon more of water. Knead the dough a couple of times with the heel of your hand until it barely holds together. (Overworking makes the dough tough—what you want is a coarse texture.) Flatten the dough into a disk, wrap in plastic wrap and chill in the refrigerator for 30 minutes.

Let the dough sit at room temperature for a short time before rolling so it will be workable.

Yield: 2 pie crusts

Raspberry Meringue Puffs

This is an elegant company dessert that you can prepare ahead. Meringues are easy to make, but they don't like humid weather—if you make them on a muggy day they won't be crisp. If using frozen berries, buy the unsweetened ones. This is also delicious with blackberries.

Meringue Puff:

6 egg whites
½ teaspoon cream of tartar
¼ teaspoon salt
1 ½ cups granulated sugar
1 teaspoon vanilla extract

Topping:

1 cup whipping cream
¼ cup confectioners' sugar
1 Tablespoon raspberry liqueur
1½ cups fresh or frozen raspberries
2 Tablespoons sugar, or to taste

Line a baking sheet with baking parchment. Preheat oven to 275°.

To make the meringue puff: in a large bowl beat the egg whites with cream of tartar and salt until foamy. Beat in the granulated sugar 2 tablespoonfuls at a time, beating well after each addition. Beat in the vanilla extract and continue beating until the meringue makes stiff, glossy peaks when you raise the beater.

Fill a pastry bag fitted with a large, round nozzle with the meringue mixture. Pipe a spiral onto the prepared baking sheet to make one large 9-inch round or a number of small rounds. Form sides of the meringue nest by piping rosettes around the edge. You can also shape the meringue with a spoon. Bake at 275° for 60 minutes. Turn off the oven but do not open the door for 4 hours, or until meringue is completely cool.

Remove the cooled meringue from the oven, place on a serving plate and let stand until needed.

To make the topping: whip the cream with the confectioners' sugar and the raspberry liqueur until stiff. Spread the cream evenly in the meringue nest and cover with raspberries which have been sugared to taste.

Yield: 8 to 10 servings

Raspberry Cream Torte

Three layers of shortbread sandwich whipped cream and berries.

1½ cups all-purpose flour
grated rind of ½ lemon
¼ cup sugar
½ cup butter
¼ cup hazelnuts or almonds, skins removed, chopped fine
1 egg yolk, beaten
3 Tablespoons ice water
4 cups fresh or frozen raspberries (thawed and drained)
2-4 Tablespoons sugar
1¼ cup whipping cream
1 Tablespoon confectioners' sugar, sifted

Whisk together flour, lemon rind, and ¼ cup sugar. Cut the butter into small pieces. Drop them into the bowl, and work with a pastry blender until the mixture resembles fine bread crumbs. Add the nuts.

In a small bowl mix beaten egg yolk with ice water. Add it to the flour mixture and mix gently with a fork until it forms a dough. Turn dough out on a floured board and knead 15 times. Form a ball, wrap in foil, and chill it in the refrigerator for 30 minutes.

Preheat oven to 350°. Divide the chilled dough into three equal parts. On a well-floured surface, roll out one part ¼ inch thick making a 9" x 3½" rectangle. Repeat with a second piece of dough. Place the rectangles on a large baking sheet 1-inch apart. Roll out the remaining dough and cut out 3-inch circles using a fluted biscuit cutter. Using a sharp knife, cut each circle in half. Place an inch apart on the baking sheet. Bake until nicely browned, about 20 to 25 minutes. Cool on a wire rack.

In a small bowl, toss the berries with sugar. Set aside.

To assemble, whip the cream with the confectioners' sugar until it forms soft peaks. Place in a piping bag fitted with a star-shaped nozzle and pipe half the cream onto one of the rectangles. (Or, spoon whipped cream onto the pastry.) Top with the fruit, reserving a few berries for decoration. Cover with the second pastry rectangle. Pipe or spoon swirls of cream down the center of the shortcake. Arrange 6 semicircles down this bed of cream. Decorate the top whole berries and serve.

Yield: 6 servings

Maple Cranberry Tart

A luscious tart of apples sautéed in butter and mixed with cranberries, nuts, maple syrup and spices.

¼ cup butter
4 tart apples, peeled and cut in thin wedges
⅔ cup maple syrup
1 cup cranberries
⅔ cup chopped pecans
1 teaspoon cinnamon
¼ teaspoon nutmeg
2 Tablespoons flour
Pastry for a 9-inch pie crust, see pages 38, 46 or 49

Melt butter in a large skillet over medium heat. Add the apples and cook them, stirring often, for 5 to 8 minutes or until almost tender. Add maple syrup, cranberries, pecans, cinnamon, and nutmeg. Cook until the cranberries pop, about 3 to 4 minutes. Remove from heat. Sprinkle flour over the mixture and gently stir it in to mix completely. Cool filling.

Preheat oven to 375°. Prepare pie crust as recipe directs. Roll out the dough and fit it into an 9-inch pie plate. Crimp the edges as desired. Pour in the filling. Bake at 375° for 25 to 30 minutes, or until the crust is golden brown.

Yield: 1 pie, about 8 servings

Dutch-Style Cranberry Apple Pie

This harvest pie brims with walnuts and cashews, light and dark raisins, sweet apples, and tart cranberries.

1 recipe pie crust, page 38
4 cups cranberries
2 cups apples, peeled, cored, and chopped
¼ cup chopped walnuts
¼ cup chopped cashews
½ cup sugar
3 Tablespoons flour
1 teaspoon cinnamon
2 Tablespoons melted butter
¼ cup light raisins
¼ cup dark raisins

Preheat the oven to 425°. Prepare pie crust as directed for a two crust pie. Line a 9-inch pie plate with the pastry. Roll out top crust and set aside.

In a large bowl mix the cranberries, apples, nuts, sugar, flour, cinnamon, butter, and raisins. Place the filling in the lined pan. Place the top crust over the filling and crimp the edges together to seal. Slash the top crust in 2 or 3 places.

Bake at 425° for 45 to 50 minutes or until well browned. (Lower the oven temperature to 350° during the last 10 minutes if the pie seems to be browning too fast.)

Yield: 1 pie

When you buy fresh cranberries look for plump, shiny ones. The color can be light or dark, depending on the variety, but it should be bright. Avoid berries that are soft or shriveled or have brown spots.

Dreamy Berry Sorbet

Ripe berries and sugar are combined with no cooking to soften the bright, fresh taste of this sorbet. The texture is dense and creamy, and the color is deep. Serve with crisp, wafer-thin butter cookies.

2 pints any berries, stems removed (about 4 cups)
¾ cup superfine granulated sugar

Wash and drain the berries. Puree them in a food processor or blender. Add the sugar, pulsing the motor of the food processor or blender. Taste and adjust sugar if needed. (Add up to ¼ cup more, depending on the sweetness of the berries.) Keep in mind that cold dulls sweetness so frozen desserts need more sugar than those eaten at room temperature.)

Pour the puree into the container of an ice cream freezer and freeze according to the manufacturer's instructions.

Yield: 1 quart

Granite de Framboise

This is a lovely, French raspberry ice, good enough for your most elegant dinners. If you want to really show off, serve it in tiny liqueur glasses between courses.

2 cups sugar
2 cups water
4 cups fresh raspberries
¼ cup lemon juice

Combine the sugar and water in a saucepan. Bring slowly to a boil, stirring until the sugar is dissolved. Boil uncovered for 5 minutes. Remove from the heat and cool.

Put the raspberries through a food mill, or puree them lightly in the food processor, taking care not to break the seeds, and then remove the seeds by pressing the puree through a strainer.

Combine raspberry puree, syrup, and lemon juice in the container of an ice cream maker and freeze according to the manufacturer's instructions.

Yield: 12 servings

Sauces, Soups & Shakes

The Berry Book

Blueberry Syrup

Serve this syrup with blueberry pancakes, French toast, or puddings. The amount of water you use will depend on the blueberries. Big, fat ones ooze more liquid into the syrup than tiny ones.

2 cups fresh or unsweetened frozen blueberries
¼ cup water, or more as needed
¾ cup sugar
½ teaspoon lemon juice

Place berries, ¼ cup water, sugar, and lemon juice in a saucepan. (Use up to ¾ cup water if your berries are tiny and not especially juicy.)

Bring to a boil. Lower heat to medium and boil uncovered until syrupy, about 20 to 30 minutes. Stir occasionally to keep the syrup from sticking. Cool and store covered in the refrigerator.

Yield: about 1 ½ cups

To store ripe raspberries, discard the moldy or squashy ones. Do not wash raspberries until just before using. They absorb water which then breaks down their flavor and texture. You can store raspberries for one or two days in the refrigerator. Place them in a single layer (they crush easily) on a paper-towel lined tray. Don't cover them and don't put them in the vegetable drawer.

Cranberries Napoleon Sauce

This fruit sauce is delicious on ice cream. I love it on rum raisin, but it goes with anything. Try it on the Dreamy Berry Sorbet, topped with a swirl of whipped cream.

> *6 ounces fresh or frozen cranberries (½ package)*
> *1 cup sugar*
> *1 cup dry white vermouth*
> *½ cup orange liqueur*
> *Juice and grated zest of ½ lemon*
> *Juice and grated zest of ½ lime*
> *1 2-inch stick of cinnamon*
> *⅛ teaspoon ground cloves*
> *⅛ teaspoon ground ginger*
> *1 can (11-ounces) mandarin oranges*

In a saucepan, combine the cranberries, sugar, vermouth, orange liqueur, juice and zest of lemon and lime, cinnamon stick, cloves, and ginger. Drain the liquid from the mandarin oranges and add ¼ cup of it to the saucepan. Add the mandarin orange segments. Bring to a boil over medium heat, stirring until the sugar is dissolved. Simmer uncovered over medium heat for 15 minutes, stirring occasionally.

Remove from heat and cool. Store in the refrigerator (will keep for several months) or freezer (will keep up to a year).

Yield: 3 to 4 cups

Cranberries grow in bogs which are kept dry while the berries grow. At harvest time the bogs are flooded. Machines shake the vines under the water, the berries come loose and float to the surface where they can be scooped up.

Blackberry Sauce

This deep purple sauce turns any simple dessert into something special. Try it over warm gingerbread, fruit turnovers, or apple pie wedges.

> 1½ cups water
> 1½ pints fresh blackberries
> 1½ cups sugar
> 2 Tablespoons cornstarch
> ⅛ teaspoon salt
> 2 teaspoons lemon juice

Place water in a 3-quart saucepan over medium heat and bring to a boil. Add berries and bring to a boil again.

In a small bowl mix sugar, cornstarch, and salt. Stir this into the berries and cook, stirring constantly, until thickened. Stir in lemon juice.

Yield: 2 ½ cups

Raspberry Sauce

This sauce makes a delicious Peach Melba dessert. Fill lightly poached peach halves with scoops of ice cream, ladle the Raspberry Sauce over the top, and garnish with toasted, slivered almonds.

> 2 pints raspberries, fresh or frozen
> ½ cup water
> ½ cup sugar
> 2 Tablespoons cornstarch
> 1 teaspoon lemon juice

Mash the berries and press them through a sieve to strain out the seeds. Place the berry puree in a saucepan with water, sugar, cornstarch, and lemon juice and stir until smooth. Cook over medium heat until thick, stirring constantly.

Yield: 3 cups

Champagne Raspberry Sauce

Ladle this wonderful sauce over fresh or poached fruit, ice cream, or rich baked custard. For a pretty dessert presentation, scoop mocha ice cream on top of sliced pound cake, then drizzle with Champagne Raspberry Sauce.

2 cups raspberry preserves
¾ cup champagne
1 Tablespoon orange zest

Place all ingredients in a small bowl and whisk until combined. Chill. Store covered in the refrigerator.

Yield: about 3 cups

Raspberry Conserve

Those who knows what's good like to heap this conserve on their toast, muffins, pancakes, and waffles. It's easy to make and keeps well in the refrigerator.

5 cups raspberries
1 cup sugar

Place raspberries in a 1-quart saucepan over medium heat, and cook them gently in their own juice until soft enough to mash easily, 1 to 2 minutes.

Press them through a food mill or a strainer to remove seeds. Extract as much pulp as possible.

Return the puree to the saucepan. Add sugar and cook over medium heat, stirring occasionally, until the sugar dissolves. Then increase heat to medium or medium-high and boil rapidly, uncovered, until thickened.

Yield: 1 cup

Cranberry and Ginger Sauce

This peppy sauce is both a marinade and a dessert sauce. Ginger gives it a warm flavor and softens the cranberry tartness just enough to be interesting. Use it as a marinade or baste for poultry, roast pork, or ribs. For a dessert treat, serve it warm over ice cream or baked apples.

¾ cup sugar
¾ cup water
2 cups cranberries
2 Tablespoons chopped crystallized ginger

In a l-quart saucepan place sugar, water, and cranberries. Bring to a boil over medium-high heat. Then lower the heat to medium and simmer uncovered for 5 minutes.

Stir in the chopped ginger. Remove the pan from the heat and let the sauce stand until cold. The sauce will keep for several weeks in the refrigerator. You can also freeze it.

To use as a marinade: marinate ribs, chicken, Cornish hens, quail or turkey overnight in the sauce. Bake uncovered at 350° until done. Use to baste roast or grilled meats and poultry during the last half hour of cooking.

Yield: 1 to 2 cups

Note: Crystallized ginger is available in the oriental section of most supermarkets or in stores where oriental groceries are sold.

Cranberry Applesauce

This version of applesauce is deep rose-red, tart, and delicious. Serve it with pancakes at breakfast or as a colorful holiday dinner side dish.

⅓ cup water
8 cups apples, peeled, cored, and sliced (about 8 medium)
4 cups fresh or frozen cranberries
1 3-inch cinnamon stick
2 whole cloves
2 cups sugar

Place water, apples, cranberries, cinnamon stick and cloves in a 3-quart saucepan over medium heat. Bring it to a boil. Reduce the heat to medium-low and simmer, uncovered, for 15 minutes or until apples are just tender. Stir in the sugar and cook 5 minutes more. Cool. Remove cinnamon stick.

Yield: 5 cups

Jazzed-Up Cranberry Relish

This is a deliciously different relish. Only the cranberries are cooked; the other ingredients are fresh and full of crunch.

2 cups cranberries
1 cup sugar
½ cup water
1½ teaspoons grated orange peel
½ cup orange juice
¼ cup light seedless raisins
¼ cup dark seedless raisins
½ cup chopped celery
½ cup walnuts, chopped medium
½ cup red grapes, halved and seeded
½ cup chopped apple
½ teaspoon ground ginger

Place cranberries, sugar, and water in a l-quart saucepan over medium heat and bring to a boil, stirring frequently. Reduce the heat to medium-low and boil gently, uncovered, for 15 minutes.

Remove from the heat and stir in the orange peel, orange juice, raisins, celery, walnuts, grapes, apple, and ginger. Pour into a jar.

When cool, cover and refrigerate.

Yield: 3 cups

Raspberry-Strawberry Freezer Jam

Because it isn't cooked, the flavor of this jam is bright and fresh and it takes just minutes to make. You can use any combination of berries to make this jam.

1 pint raspberries, crushed (about 2 cups)
1 pint strawberries, crushed (about 2 cups)
4 cups sugar
¾ cup water
1 package powdered fruit pectin

Mix crushed fruit and sugar in a large bowl and let stand for 10 minutes.

In a small saucepan mix water and fruit pectin. Bring the mixture to a boil and boil for 1 minute, stirring constantly.

Remove from the heat and stir into the crushed fruit. Continue to stir for 3 minutes.

Ladle the jam into six 1-cup containers. Cover immediately with tight lids. Let stand at room temperature for 24 hours. Then store in the freezer.

Yield: 6 cups jam

A few different ways to enjoy berries: The Italians pour a few drops of balsamic vinegar over fresh berries and eat them for dessert. The Spanish pour on a few spoonfuls of fresh orange juice. Some people grind black peppercorns over strawberries. Before you wrinkle your nose, try it—you might be surprised.

Strawberry Soup

Serve as a delightful first course or as a light dessert. You'll love the deep strawberry flavor with a whisper of wine, a hint of lemon, and a rich ruby color. Make it in minutes and enjoy it at leisure.

1 pint fresh or frozen unsweetened strawberries (2 cups)
½ cup dry white wine
1 teaspoon grated lemon peel
½ cup sugar
1 Tablespoon lemon juice

Hull the strawberries. Wash, rinse, and dry them. Slice 3 of the berries and set them aside for garnish.

Place the remaining berries in the blender or processor bowl along with the wine, lemon peel, sugar, and lemon juice and puree until smooth. Pour into a bowl and refrigerate until well chilled.

Serve in chilled soup bowls garnished with the reserved sliced berries.

Yield: Serves 4

Raspberry Port Wine Soup

Port wine turns this jewel-colored soup into a sophisticated dish.

2 packages (10-ounces) frozen raspberries, thawed
1 cup light or dark port wine
2-inch stick of cinnamon
2 teaspoons cornstarch
¼ cup water
mint, as a garnish

Place raspberries, port, and cinnamon stick in a saucepan and bring to a boil. Simmer uncovered over low heat for 10 minutes. Mix cornstarch with water and stir into the soup. Cook and stir until thickened.

Pour into a bowl and chill in the refrigerator. Just before serving, remove the cinnamon stick. Garnish with mint.

Yield: Serves 4

Raspberry Lemonade

This is refreshing on a hot, summer day.

1 cup granulated sugar
11 cups water
1 cup freshly squeezed lemon juice
½ cup crushed raspberries
1 unpeeled lemon, sliced

Mix sugar with 2 cups of the water in a small saucepan. Heat until the sugar dissolves. Cool and pour into a 4-quart bowl. Add lemon juice, raspberries, and remaining water. Chill well. Serve in a punch bowl or pitchers. Garnish with lemon slices.

Yield: Serves 12

Berry Banana Frost

This is a nutritious way to start the day.

1 cup berries (any variety)
½ banana, cut into chunks
1 cup milk, skim or whole
½ teaspoon vanilla or almond extract
¾ cup ice

Place all ingredients in the container of a blender or food processor. Blend until smooth. Serve immediately.

Yield: 1 drink